Thomas Churchyard

A light bondell of liuly discourses called Churchyardes charge

Presented as a Newe yeres gifte to the Earle of Surrie

.

Thomas Churchyard

A light bondell of liuly discourses called Churchyardes charge
Presented as a Newe yeres gifte to the Earle of Surrie

ISBN/EAN: 9783337197360

Printed in Europe, USA, Canada, Australia, Japan

Cover: Foto ©berggeist007 / pixelio.de

More available books at **www.hansebooks.com**

A light Bondell of li-
uly diſcourſes called Churchyardes Charge, preſented as a
Newe yeres gifte to the right honou
rable, the Earle of Surrie, in whiche
Bondell of verſes is ſutche varietie of
matter, and ſeuerall inuentions, that
maie bee as delitefull to the Reader,
as it was a Charge and labour to the
writer, ſette forthe for a peece
of paſtime, by *Thomas*
Churchyarde
Gent.

¶ Jmprinted at London,
by *Ihon Kyngston.*

1580.

Churchards Armes.

IN·DIEV· ET·MON ROY·

I Knowe not my good lorde, whether my boldnesse and presum-
ption be greater, then the base-
nesse of my matter herein pen-
ned, and I mynde to presente:
consideryng the worthinesse of
the personage, to whom I dedi-
cate my booke, and the weakenesse of my wit, that pre-
senteth vaine verses, where vertue of the mynde
aboundeth. But for that I treate not of mere trifles,
(nor meane to corrupt sound senses, and good maners
with wanton wordes or leude rime) I am partly per-
swaded this myne Newyeres gift, shall giue your lord-
ship delite, and purchace to my self the desired thankes, that euery honest writer deserueth: Because the
substance and effect of all my inuentions, are shadowed
vnder the sheld of good meanyng: And a matter well
meant (by the courtesie of true construction) maie passe

*.ij.   the

The Epiſtle

the muſter & good opinion of the people, emong the beſt
aſſemblies that looketh on the furniture J bryng, and
ablenesſe of my penne. And albeit ſome weapons want
to beate backe the thompyng boltes of euill tongues (in
my defence be it ſpoken) yet the Armour of right, and
Target of trothe ſhall bee ſufficient to ſtrike doune the
blowes, that hautie hartes with threatnyng thwartes
can offer . And who ſo euer haſtely or vnaduiſedly
through malicious wordes , hinders the credite of any
honeſt workes, maie be thought both a raſhe and a par-
ciall ſpeaker, & a buſie medler in matters, thei neither
mynde to amende, nor nor will ſuffer that the worlde
ſhall ſpeake well therof. But now farther to procede, &
enter into the cauſe of this my boldneſſe, the troth is in
callyng to remēbrance a promes that I made, touching
ſome verſes. And honoryng in harte the Erle of Surrie, your Lordſhipps graundfather, & my maſter (who
was a noble warriour, an eloquent Oratour, and a ſecond Petrarke) l could doe no leſſe but publiſhe to the
worlde ſomewhat that ſhould ſhewe, I had loſt no time
in his ſeruice. And finding an other of his race and to-
wardneſſe, who hath taſte and feelyng in the good gif-
tes of Nature, and noble vertues of his aunceſtours,
(the hope of whiche graces, promiſeth greate perfecti-
on to followe in tyme to come) J thought J might de-
dedicate

Dedicatorie.

dicate a booke vnto your Lordſhippe, named by myne
owne liking Churchyards Charge. But now right noble Earle, the worlde louyng change and varietie of
matter, waxeth awearie of freuoulous verſes (becauſe
ſo many are writers of Mieter) and looketh for ſome
learned diſcourſe, by whiche meanes my barrain bookes maie remaine vnred, or miſliked, and ſo lye on the
Stationers ſtall, as a ſillie ſigne of a newe nothyng, neither worthe the buiyng, nor the regardyng. To that I
maie aunſwere (vnder pardon and correction) that
the grounde whiche of Nature yeldeth, but Thiſtles or
Brambles, maie bryng forthe no good Corne of it ſelf,
contrary to his operation and kinde. Nor a man that
is accuſtomed to treate of trifles maie, not meddle
with the deepeneſſe of graue argumentes. For as it paſ-
ſes the ſearche and capacitie of a ſimple witte, to ſe into any matter of importaunce, ſo it is neceſſarie that a
pleaſaunt and plaine companion, ſhould alwaies be occupied about paſtymes, and namely at Chriſtmas, whē
little ſhort tales, driues out a pece of the long nighres,
and rather with mirthe to procure a laughter, then
with ſadneſſe prouoke a lowryng: and he that ſturreth
vp the heauie myndes to lightſome conſaites, is more
welcome in euery place, then he that ouerthrowes the
weake ſenſes of common people, with curious imagina-

*.iij.   tions,

tions, and burthens bothe bodie and mynde, with wordes of greater weight, then common iudgement can cōceiue, and be able to beare. A tale or a toye mirrely deliuered, pleaſeth moſte mennes eares : and an earneſte ſadde argument, either rockes a man a ſlepe, or maketh the hearers awearie. And the nature of Rime is to reuiue the ſpirites, or moue a ſmile, when many a one is ſcarce pleaſauntly diſpoſed. A Rime goeth on ſutche feete, ſtandeth on ſutche iayntes, and rappeth out ſutch reaſons, that wiſedome taketh pleaſure in, and follie will make a wonder of. The woordes by inuention hits a thyng ſo iompe, and kepes ſutche a decorum and methode, that bothe order and meaſure is ſeen, in the cunnyng conueyance of the verſes, eſpecially if the ſwete and ſmothe ſentences bee ſifted, from the ſowre rough Branne of needeleſſe babble and vanitie · A ſenſible witte hauyng the pennyng of the matter. But loe my good Lorde, in ſhewyng the nature and qualitie of a good verſe, how my hoblyng is ſeen, and perceiued by the badneſſe, or bare handelyng of the thynges herein written: yet now J haue ron ſo farre in ouerweenyng, that either J am forced to goe forward, or remaine in the midwaie diſcomforted, and without remeadye. VVherefore, albeit J ſhall ſhewe but a bondell of drie deuiſes, J muſt open my fardell, & make ſale of ſutche

ſtuffe,

Dedicatorie.

ſtuffe, as my hedde hath been ſtuffed withall: Euen like the poore Peddlar, that trudgeth with his packe to a Faire, and there vnfoldeth emong ſome newe laces & odde trifles, a greate deale of old ware and little rē-nantes, that for lacke of quicke ſale, hath laine long in a cloſe corner. I neede not to ſeeke out a patron to ſupport them: for thei are neither worthe the readyng, nor the buiyng, yet hauyng no better, am compelled to vtter the thynges I haue lefte. Prouidyng that my nexte booke maie ſhewe ſomewhat emong the reſt that goeth before: for that it ſhall be dedicated to the moſte worthieſt (and towardes noble man), the Erle of Oxford, as my laiſure maie ſerue, and yet with greate expedition. Thus beyng ouer tedious and bolde, in ſtretching out a ſhort and ſorie Epiſtle (that had been better knit vp in fewe lines) I wiſhe your Lordſhip many newe and happie yeres, long life to your liking, to the honour of GOD, and encreaſe of good fame: and a peece or portion of eche goodneſſe can be named.
From my lodging nere to the Courte the
firſt daie of Ianuarie. Your Lord-
ſhippes alwaies at com-
maundemente.

Thomas Churchyard.

To the freendly Reader.

Daily trouble the good Reader with Bookes, Verſes, Pamflettes, and many other tri- flyng thinges, as mutche to hold thee occupied in good will to- wardes me (and keepe thee from loſſe of tyme) as for any matter that I either cã gaine glorie by, or deſerueth to bee embraſed: but vſyng me and my workes thankefully, and paiyng me for my pai- nes, with the like courteſie, that other men reapeth for their labours , I thinke my ſtudie well beſtowed, and promes yet with my penne , to pleaſure thee farther. And for that I would haue all menne to thinke, that in trothe and plainneſſe I haue greate felicitie, and doe hate any kinde of flatterie or fineneſſe. I meane in my next booke, called my Challenge, to ronne ouer ma- ny of myne other woorkes, and where peraduenture by ſome reporte of others, (that knewe not the trothe,) I haue failed in ſettyng foorthe of ſome ſeruices, emong the whiche Maiſter Jhon Norrice, and diuers wor-

The Preface.

thie gentlemen *Captaines* now in *Flaunders*, haue not the worthinesse of their seruices declared. *J* doe promes that now beyng better instructed, and hauyng true intelligence of thynges as thei were, *J* will at large write the commendation of as many, as merites to bee honoured for their well doyng, and make amendes, where either by ignoraunce, or the report of others *J* haue failed. For so sure as GOD is *Almightie*, if *J* could gaine mountaines of golde, to flatter any one in *Printyng* an vntrothe, *J* would rather wishe my handes were of, then take in hande sutche a matter. For neither affection, fauor, commoditie, fame, nor parciallitie, at no tyme nor season, shall willyngly lead my penne amisse. And farther, if *J* thought any one for his owne glories sake, had tolde me more then is truely to bee proued : *J* would not onely condempne my self for giuyng sutche hastie credite, to vainglorious people : But in like sorte my penne should shewe the blotte in their browes, that giueth me wrong aduertisementes. And so good Reader, condempne not me if anything bee amisse, or lefte out that ought to haue been touched : For as *J* knowe and am instructed) seeyng not all my self) *J* must write, and so till my nexte booke come forthe (where many thynges shall be treated of) and that my good will to the honouryng of vertue

shall

The Preface.

shall bee seen. I bidde thee fare well freendly Reader, crauyng thy fauourable Iudgemente on that I haue written. From those men for whom my booke hath bin a blast of fame vnto (as I did beleue by the true trompet of penne) are not of sutche greate abilitie, that either their purses power or countenaunces, could comptll me to followe their humours: nor by any meanes woorke my muse to their willes, whose ritchesse and wealthe is not able (if men would be hired) to wrest a wrie the hande and hedde of an honest writer. And to make manifest that I neither willbe infected, nor carried awaie from that whiche is good, for any badde practice and perswasions, I confesse before GOD and the worlde, I scarce haue receiued thankes, for the honest labors I haue taken (at their handes that I haue written of) mutche lesse haue I been recompenced, or founde cause to flatter the worlde. But in one thyng I maie reioyce, the honourable persone to whom my choice is dedicated (and others of great callyng) hath bothe been gratefull sondrie waies (in moste bountifull maner:) and also hath encoraged me to proceede in the like paines, whiche in very deede I mynde to go about as well to the fame and glory of good menne, as for the aduoidyng of sloth and idelnesse my self.

FINIS.

¶ A storie translated out of Frenche.

IN old tyme paste in Picardie,
 there dwelt an honest man,
Whose name the storie doeth not tell,
 what he was called than:
A wife he had, a house he helde,
 as Farmers vse to doo,
And lacked little for thesame,
 that did belong there too.
And as God sent hym suffisaunce,
to rubbe forthe life here lent,
So for to chere vnweldie age, faire children God hym sent:
Of whiche he had one moste in minde, a lad of liuely spreete,
Who with great care he kept to schoole, as for his youth was meete.
This boye to glad his fathers harte, in bookes set his delite,
And learnd to make a Latine verse, to reade and eke to write:
And for his Nature was enclinde, to studie learnyngs lore,
The better he aplied his schoole, he profited the more.
To make his schoole the sweeter seem, with Musicke mixed was,
The studie that he followd then, the tyme awaie to passe: '(smal let,
Good bookes were bought and instruments, greate charge was but
If that thereby the father might, the sonne some knowledge get.
In seuen yeres (as tyme it was,) this striplyng gan to taste,
Tyme well emploied, tyme driuen forthe, and tyme ill spent in waste,
And made no small account thereof, but still sought more to haue,
Wherewith he to his father came, on knees this did he craue.
I haue ꝙ he dere father now, my childishe daies ore rome,
And as I thinke, and you beleue, my boyes delites are donne:
And as my witte and grace hath seru'd, some learnyng haue I gote,
And as I knowe you loue me well, on me you should not dote.
I meane I should not still at home, vnder my mothers wing,

Be brought vp like a wanton child, and doe no other thing:
The worlde is wide, I want no witte, your wealth is not so greate,
But you maie thinke in some dere yere, I scarce deserue my meate.
And though your kinde and custome is, full fatherlike alwaie,
Yet should your sonne discretion haue, to ease you as he maie:
Wherefore to make your burthen lesse, let me goe seeke my happ,
And let no longer now your sonne, be lullde in mothers lapp.
The father wise well vnderstoode, his childs request at full,
And that the setters of his youth, he thought awaie to pull:
(Before he gaue hym leaue to parte) by councell graue and sage,
Well boye quoth he now art thou come, vnto thy flowryng age.
Now art thou like the little wande, that bent and bound will bee,
Unto his hande or to his skill, that liste to maister thee:
Now are ripe peres soone rotten made, now art thou apt to take,
Bothe good and badd, but cheefly things, that age bidds thee forsake.
And now large scope shall sone forget, what short rein learnd in schole
And thou that wisely wast brought vp, shall plaie the wanton foole.
Abrode as wilde harebrains are wont, newe taken from their booke,
And in a while laie all a side, nere after their on looke.
In eury place of thy repaire, thou shalt no father finde,
Nor scarce a freende to whom thou maiest, at all tymes sho thy minde:
But on Gods blessyng goe thy waie, thy wilde Otes are vnsowne,
Hereafter time shall learne thee well, things to thee now vnknowne.
The ladde his leaue and farewell tooke, well furnisht for the nonce,
And had about hym as I trowe, his treasure all at once:
To court he came all maisterlesse, and sawe what likt hym beste,
Of runnyng Leather were his shues, his feete no where could reste:
His book es to blade and bucklar chang'd, he gaue ore scholars trade,
Where reuell roysted all in ruffe, there he his residence made.
This rule had soone his purse so pickt, that princoks wanted pence,
And oft he sawe some trussed vp, that made but small offence:
His father farre from seyng this, he come of honest stocke,
He hofsprong forthe a hatefull life, in many a wicked flocke.
And prickved oft to slipper shifts, yet some regard he tooke,
To be a sclander to his kinne, that kept hym to his booke:
And in a better moode to thriue, to seruice did he drawe,

He

He muſt goe that the deuill driues, ye knowe neede hath no lawe.
A maiſter of no meane eſtate, a mirrour in thoſe daies,
His happie Fortune then hym gate, whoſe vertues muſt I praiſe:
More heauenly were thoſe gifts he had, then yearthly was his forme.
His corps to worthie for the graue, his fleſhe no meate for wormе.
An Erle of birthe, a God of ſprite, a *Tulle* for his tong,　(rong:
Me thinke of right the worlde ſhould ſhake, when half his praiſe were
Oh curſed are thoſe crooked crafts, that his owne countrey wrought.
To chop of ſutche a choſen hed, as our tyme nere forthe brought.
His knowledge crept beyond the ſtarrs, & raught to Ioues hie trone
The bowels of the yearth he ſawe, in his deepe breaſt vnknowne:
His witt lookt through eche mās deuice, his iudgemēt groūded was,
Almoſte he had foreſight to knowe, ere things ſhould come to paſſe.
When thei ſhould fall what ſhould betied, oh what a loſſe of weight,
Was it to loſe ſo ripe a hedde, that reached ſutche a height:
In eury art he feelyng had, with penne paſt *Petrarke* ſure,
A faſhon framde whiche could his foes, to freendſhip oft alure.
His vertues could not kepe hym here, but rather wrought his harms,
And made his enemies murmure oft, & brought them in by ſwarms:
Whoſe practiſe put hym to his plonge, and loſte his life thereby,
Oh cancred breſts that haue ſutche harts, wherin ſuche hate doth lye.
As told I haue, this yong man ſeru'd, this maiſter twiſe twoo yere,
And learnd therein ſutche fruitfull ſkill, as long he held full dere:
And vſd the penne as he was taught, and other gifts alſo,
Whiche made hym hold the capp on hed, where ſome do croch full lo.
As credite came he carefull was, how to maintaine the ſame,
And made ſmall count of life or death, to kepe his honeſt name:
His father not a little glad, of his good happ thus founde,
And he forgot no duetie ſure, to whom he ought be bounde.
From court to warrs he wounde about, a Soldiours life to leade,
And leaned to the worthieſt ſort, their ſtepps to marche and treade:
And followd Camon wheele as faſt, to learne ſome knowlege then,
As he afore at maiſters heeles, did waite with ſeruyng men.
But thoſe twoo liues a diffrence haue, at home good chere he had,
Abroad full many a hongrie meale, and lodgyng verie bad:
All daie in conſlict caled faſte, whiche made his ſhulders ake,

Churchyardes Charge.

All night vpon a couche of ſtrawe, right glad his reſt to take.
Through thicke & thin a thriftleſſe tyme, he ſpent & felt mutch gréef,
And euer hopyng for the ſame, at length to finde reléef:
No ſmall while there as ye haue hard, in colde ſharpe winter nights,
Where he did feele ſtrange plags enowe, and ſawe full vgly ſights.
Some dy for lack, ſome ſeke for death, ſome liue as though ther were
Ne God nor man, nor torment here, or hence we ought to feare:
But yet he markt ſome of that ſort, whoſe eſtimation ſtood,
Vpon eche point of honeſt name, and things that ſemed good.
He ſawe likewiſe how Fortune plaied, with ſome men for a while,
And after paid them home for all, and ſo did them begile:
A wearie of theſe waſtyng woes, a while he left the warre,
And for deſire to learne the tongs, he trauelo very farre.
And had of cury langage part, when homeware did he drawe,
And could rehearſall make full well, of that abroad he ſawe:
To ſtudie wholie was he bent, but countreis cauſe would not,
But he ſhould haunt the warrs againe, aſſigned thereto by lot.
And eke by hope and all vaine happ, procured to the ſame,
As though eche other glorie grewe, on warrs and warlike fame:
Without the whiche no worlds renowme, was worth a flye he heeld,
For that is honour wonne in deede, once got within the feeld.
Thus in his hed and hye conſaite, he iudg'd that beſte of all,
And thought no mouth for Suger mete, that could not taſte the gall:
Good lucke and bad mixt in one cup, he dranke to quenche his thirſte,
And better brookt the ſecond warrs, then he did like the firſte.
And leſſe found fault in fortunes freaks, time had ſo well him taught
At chances ſowne he chang'd no chere, nor at ſwete haps much laught:
In priſon thriſe, in danger oft, bothe hurt and mangled ſore,
And all in ſeruice of his prince, and all awaie he wore.
In meane eſtate in office too, ſometyme a ſingle paie,
Some tyme fewe had ſo muche a weeke, as he was loude a daie:
When worlde waxt wiſe, & wealthe did faile & princes pride appald,
And emptie purſe, and priuie plag's, for perfite peace had cald.
And kings and kingdoms quiet were, this man to court he came,
Newe from the giues with face and lookes, as ſimple as a Lame:
Freſhe from his enemies hands came he, where for his countries right,

He

He priſned was and forſte to grant, a ranſome paſt his might,
Sent home vpon a bande and ſeale, whiche is to ſtrange a trade,
There to remaine till he for helpe, ſome honeſt ſhifte had made:
All ſpoiled cleane bare as the bird, whoſe feathers pluckt haue bin,
Bothe ſicke and weake his colour gon, with cheeks full pale and thin.
The ſight ſo ſtrange or worlde ſo nought, or God would haue it ſo,
This man had ſcarce a welcome home, whiche made him muſe I tro
His countrey not as he it left, all changed was the ſtate,
But all one thing this man deſeru'd, therein no cauſe of hate.
A careleſſe looke on hym thei caſte, ſauyng a fewe in deede,
Through warrs brought lowe for ſeruice ſake, & felt therby his neede
Of ſuche as could a diffrence make, of drom and trompetts ſounde,
(From tabber pipe & Maipole mirth,) their helpyng hands he founde:
And thoſe that fauord featts of warre, and ſauour tooke therein,
With open armes embraſte hym hard, and ſaid where haſt thou bin.
But none of theſe could doe hym good, to ſet hym vp I meane,
His freends decaied his father dedde, and houſholde broke vp cleane:
Craue could he not, his hart ſo hye, it would not ſtoupe to ſteale,
He ſcornde to ſerue a forraine prince, prefarryng common weale.
Aboue all other things on earth, his countrey honourd he,
At home he likt more poore eſtate, then thence a lorde to be:
Where ſhould he ſue where ran thoſe ſprings, could cole his feuer hot,
Where durſt he mone or plaine for ſhame, where might relief begot:
But at the fountain or well hedde, yea at his Princes hande,
And in a fewe well couched lines, to make her vnderſtande:
His caſe his ſcourge, loe ſo he did, and boldly did he tell,
The ſame hym ſelf vnto the Prince, who knowes the man full well.
And gracious words thrie tymes he gate, the fourth to tell you plain
Unfruitfull was things were ſtraite laeſt, faire woords maks fooles full fain:
When prince nor countrey made no count, of hym nor of his caſe,
And none of bothe would help hym home, of whom he ſought for grace.
For whom and for their cauſe alone, in enemies hands he fell,
And for their right to warrs he went, as all men knowes full well:
And loſte his blood for their defence, and for their quarell fought,
And for the ſame full ſlenderly, lookte to and ſet at nought.
When he his duetie to his powre, did eury daie and yere,

Suche vnkinde gwerdon had receiu'd, as well before you here:
He faied let *Marcus Regulus* in faine of Romains ftande,
Whiche kept his othe and did retourne, againe to Carthage lande.
If *Tullie* were a liue to write, his praifes more at full,
Yet fince I fcapt my enmies hands, at home abide I wull:
He fhould not me perfwade to goe, where nought but death is found,
My countrey cares not for my life, then why fhould I be bound.
To toies or any other bande, that I haue power to breake,
Whiche I was forced by my foe, in perfone for to fpeake:
And for the hope of countries helpe, and freends that there I had,
In any fort to pleafe my foes, I was bothe faine and glad.
Not mindyng if my countrey would, releafe me from his hande,
To breake good order any whitte, or violate my bande:
For iuftice bids eche man doe right, which God doeth know I ment,
But now a captiue peeld my felf, it maie not me content.
For where that *Tullie* doeth affirme, men ought t kepe their othe,
Unto their freends in eury point, and to their enmies bothe:
And bryngeth *Marcus Regulus*, example for the fame,
With other reafons many a one, whiche were too long to name.
He fhewed that the Senate all, would hym haue ftaied at Rome,
And as in counfaill then thei facte, their iugement and their doome.
Was that the prifners fhould be free, whiche thei of Carthage held,
And he fhould ftaye, full oft his freends, this tale to hym thei teld;
Thei proffred helpe, and offred ftill, this *Marcus* to redeeme,
But *Marcus* for a further fkill, did little that efteeme.
I finde no fuccour hope nor aide, then bounde why fhould I be,
More to my countrey in this cafe, that countrey is to me:
Thefe wordes this heauie man rehearft, fo bade the warrs *adue*,
And thought he would no ranfome paie, for any thyng he knewe.
Wherefore from court he tournd his face, and fo an othe he fwore,
As long as he his fiue witts had, to come in court no more:
He kept that othe and cut his cote, as clothe and meafure wold,
And doune to Picardie he comes, fome faied at thirtie yere old.
And for his lands and rents were fmall, a maifter lent he too,
Who vfd his feruaunt not fo well, as maifters ought to doo:
He was not made out of that mould, that his lafte maifter was,

Thefe

These twoo in vertues were as like, as Gold was vnto Glasse.
Vpon a daie alone he satte, and saied these woꝛds right sadd,
Are soldiours cast at carts arse now, that long faire woꝛds haue had:
Shall kyngs nere neede foꝛ helpe againe, is foꝛtune so their freende,
Haue thei a pattent of the Gods, this peace shall neuer ende.

God gꝛaunt yet will I shift I trowe, foꝛ on oꝛ happ shall faile,
And in the stoꝛmes my ship shall learne, to beare a quiet saile:
And cleane foꝛget bꝛaue daies agoe, that fed my youthfull peres,
Full glad that I haue gotten home, and scapt the scrattyng Bꝛeers.
Of warrs and other woꝛldly toiles, adue I see their fine,
A wife shall now content my mynde, suche as the Gods assigne:
A woo yng thus this haplesse man, rode foꝛthe not set to sale,
Though none like hym in this his suite, was meete to tell his tale.
And as the heauens had agreed, the Planetts well were bent,
Oꝛ sone descended from his hoꝛse, and boldly in he went.
Where dwelt a sober widdowe then, bothe wise and wifly too,
Late fallen sicke, vnknowne to hym, that tyme vnfitt to woo:
But her discretion was so greate, and his behauiour bothe,
These straungers fell acqueinted thus, if ye will knowe the trothe.
He faind an other errid to make, dissembling yet a space,
Till he might spie a better tyme, to shewe her all his care:
So takyng leaue foꝛ freends he wꝛought, to bꝛyng this thing about,
In suche affaires some spake full faire, that are full well to doubt.
Foꝛ commonly men take no cars, of others sutes foꝛ why,
Their pꝛofite as thei gesse themselfes, in hindꝛyng that maie ly:
Some pꝛomise helpe and see no gaine, maye spꝛing to them thereof,
Ware cold and slowe foꝛ lacke of spaꝛre, and bꝛe it as a scoffe.
An other soꝛt with stingyng tongs, saie maistres take good heede,
This man will sone your feathers pull, and cast ye of at neede:
Will you that haue bothe wealth and ease, to pooꝛ mens curse stand,
And let an other maister be, of that is in your hande.
Some seekyng rule of that she hath, and fleecyng from her first,
Doe fawne and flatter all the daie, and guide her as thei liste:
And liue on her, and hate her life, and waite her death to see,
And well can please her while she liues, her sectoꝛs foꝛ to be.
Suche instruments these widdowes haue, about them euery howꝛe,

Perchance this man perçeiu'd the like, and had good cause to lowre:
But as he knewe the fatall chance, of things comes from aboue;
So he began and sought to knowe, the fine of all his loue.
And found a daie full apt therefore, at large thesame he told,
And flatly this her aunswere was, she neuer marrie would:
If no newe thoughts fell in her minde, whereof no doubt she made,
Except she chose a wealthie man, that had a grounded trade.
To liue and had a hourd of gold, to keepe them bothe from dette,
Good sir quod she on riches sure, my minde is fully sette:
I can with ritches vertues make, vertue with want is bare,
I praie you come no more at me, thus answerd now ye are.
I would be lothe to hold you on, with wordes and meane in deede,
That neither you for all your sute, nor any yet shall speede:
He hearyng this hangde downe the hedde; and smilde to cloke his woe
A worde or twoo he after spake, and parted euen so.
The waie he rode, he curst hym self, for cruell death he cried,
And saied oh wretche thou liuest to long, to long here doest thou bide:
Not onely for this froward happ, but for all other chance,
At any tyme thou tookst in hande, thy self for to aduaunce.
Thy vertues ought if thei maie be, serues thee no whit at all,
Thy learnyng stands thee in no steede, thy trauell helps as small:
Thy knowledge sought in warrs abroad, at home doth thee no good,
Thy langage is but laught at here, where some would sucke thy blood
Thy Poetts vaine and gift of penne, that pleasurde thousandes long,
Hath now enough to doe to make, of thee a wofull song:
Thy freendes that long a winnyng were, in court and countrey plain,
Doeth serue thee to as good a ende, as mirth doeth sicke mans pain.
Thy youth though part be left behinde, whose course yet is to ronne,
With bragge of showe or seemly shape, what botie hath it wonne:
Thy honest life or manly harte, that though eche storme hath paste,
Thy reputation hardly wonne, what helps thee now at laste.
Thus to his chamber in his heate, he comes with fomyng mouthe,
And in his bloodie breast he felt, full many fitts vncouthe:
And on the bedde he laied hym downe, and for his Lute he raught,
And brake a twoo those giltlesse strings, as he had bin bestraught.
And ere he flang it to the walls, my plaiefeere fare thou well;

Saied

Saide he as sweete as *Orpheus* harpe, that wan his wife from hell:
You Instruments eche one of you, keepe well your care of woode,
And to the scrawlyng catyng wormes, J you bequeath as foode.
Up stept he to his studie doore, all that stoode in his waie
He brake and burnt bothe booke and scrowll, and made a foule araie:
Some authours saie that could not be, his wisedome did asswage,
The inward passions of his minde, and heate of all his rage.
But well J wotte he did prepare, to part from freends and all,
And staied but till the Spryng came on, for leafe was at the fall:
Now all these stormes and tempests past, this man had sutche a vaine,
When matter mou'd, and cause requierd, he went to warrs againe.
And studyng Fortune all a like, as haplesse people doe,
He fell straighewaies in enmies hands, and was sore wounded too:
But taken prisnar, promesd mutche, though little had too paie,
(A subtell shift to saue the life, and scape a bloody fraie.)
Yet still because he gallant was, and had some charge of men,
He held vp heade, and in strange place, tooke mutche vpon hym then:
The enmie seyng this yong man, bothe well brought vp and trainde,
As one that kept sutche state and grace, as he deceipt disdainde.
And to be plaine (in eury point) vpon sutche termes he stoode,
As his dissent and offspryng came, of hie and noble bloode:
Of gentill race he might make boste, but of so greate a stocke,
He could not vaunt for that deuice, was but a scorne and mocke.
Well by this meanes he was so likt, and made of eury where,
That all that laude rang of the fame, and brute that he did bere:
And so the Princes of that realme, to court did call hym tho,
Where he with feasts and triumphs greate, and many a courtly sho.
Past of the tyme, and grewe so farre, in fauour with the beste,
That he would plaie at Dice and Cards, and so set vp his reste:
For he had money when he would, and went so gaie and braue,
On credite that he finely wan, as mutche as he could craue.
And when to takers house againe, this prisner should repaire,
The greatest lords of all that soile, when he would take the aire:
Would in a maner waite at hande, to doe this prisner ease,
And well were thei of all degrees, that best this man could pleafe.
A nomber of his nation then, of right greate wealthe and state,

By this mans worde & onely band, ſtraight waie their fredome gate,
For he was bounde for eury one, that taken were before,
And ſo did for their raunſome lye, and running on the ſcore.
And brauyng out the matter through, a Ladie of greate race,
In honeſt ſort, and frenoly meane, his ſreendſhip did embrace:
Who promiſed hym, to ſet hym free, and helpe hym thence in haſte,
But ſtill about this priſner loe, a priuie gard was plaſte.
Pea ſutche a bande and daiely watche, as he might not diſceiue,
Pet he had hope in ſpite to ſcape, awaie without their leaue:
And ſhapt to flye, and giue the ſlipp, if Fortune would agree,
The watche and ward, ſhould be begulde, and priſner ſhould goe free.
And as theſe things a doyng were, a man of mutche renowne,
Was taken after in the feeld, and brought ſo to the toune:
Where hearyng of this other wight, was aſkte if he did knowe,
The ſonner perſone namde before, that daiely brau'd it ſo.
He is quod he that laſt was caught, a luſtie Souldiour ſure,
A man that mutche hath felt of woe, and greate things can endure:
Of geutill blood and maners bothe, and wants but wealth alone;
What what ſir knight, haue you ſaied trothe, and is he ſuche a one,
Then ſhall he bye his brauery dere, and paie therefore ſo well,
He ſhall not boſte of that he gains, in heauen nor in hell:
So all in ſuerie flang he forthe, and to this man he goes,
That was in deede ſo farre in debt, for meate for drinke and cloſe,
And thruſt hym in a priſon ſtrong, where feeble foode he had,
And heauie Irons whiche might make, a ſillie ſoule full ſad:
His miſtres knowyng of the caſe, her promes thought to kepe,
So wakyng in a Mooneſhine night, when neighbours were a ſlepe,
She came her nere the priſon doore, and at a windowe pried,
Where planly full before her vewe, her ſeruaunt had ſhe ſpied:
To whom ſhe ſpake and told her mynde, as cloſely as ſhe might,
And gaue hym councell in good tyme, to ſteale awaie by night.
And left hym files to ſette hym free, and robes to doe hym good,
With ſome hard egges and bread in bagg, and told hym nere a wood:
There was a broine, where ſhe would wait, for him whe time drue on
That doen ſhe tokg a frendoly leaue, for then ſhe muſt be gon.
The priſner did diſcuſſe his beſte, and bent to doe or dye,

Prepared eche thing in order well, as he on strawe did lye:
The tyme approche, of his *adue*, and she was come in deede,
Unto the place appointed right, with gold and wealth for neede.
But breaking downe a rotten wall, the prisuer was in feare,
For out of bedde his keeper stept, and asked who was there:
With that the prisuer stumbled on, a hatchet sharpe and keen,
And raught the gealer suche a blowe, that long was felt and seen.
He cried and rored like a bull, where at the village throwe,
Was vp and streight to horsebacke went, but loe the prisuer nowe:
Was at the wood, where he had found, his mistres all alone,
Who wept and blubberd like a chila, and made so greate a mone.
For that thei bothe in daunger were, but what should more be saied,
The man pluckt vp his harte and sprites, the woman sore afraied:
Ran home againe to fathers house, and he that now was free,
Had neither minde on gold nor gilt, but to the Brome goes he.
And there abode a happie howre, yea twoo daies long at least,
He lape as close on cold bare ground, as bird doeth in warme neast:
His mistres well escaped home, and in the house she was,
Before the crie and *larum* rose, so blamelesse did she passe.
And her poore seruaunt, had wide worlde, to walke in now at will,
Although he was in hazard greate, and long in daunger still:
For he had three score mile to goe, emong his enmies all,
Whiche he did trudge in foule darke nightes, and so as h app did fall.
He scape a scourge and scourging bothe, and came where he desierd,
And finely had deceiu'd his foes, what could be more requierd:
Yet long at home he could not rest, to warrs againe he went,
Where in greate seruice sondrie tymes, but half a yere he spent.
And loe his Destinie was so straunge, he taken was againe,
And clapt vp closely for a spie, and there to tell you plaine:
He was condemde to lose his hedde, no other hope he sawe,
The daie drewe on of his dispatche, to dye by Marciall lawe.
The people swarming in the streats, and scaffold readie there,
A noble Dame, his respite crau'd, and spake for hym so feare:
That then the maister of the Campe, his honest answere hard,
For whiche he came in credite streight, and was at length preferd.
To right good roome and wages too, then ritchly home he drewe,

Churchyardes Charge.

And left the warrs, and in greate heate, he for a wife did sewe,
But haste makes waste, an old prouerbe, *for he was wiud in deede,*
God sende all Souldiours in their age, some better lucke at neede:
Now he bethought hym on the woords, the widdowe tolde hym of,
Whiche long he held but as a ieast, a scorne and merrie scoffe.
She saied that witte and wealth were good, but who a wiuyng goes
Must needs be sure of wealth before, els he his sute shall lose:
For want but breeds mislikyng still, and wit will weaue but woe,
(In louers lomes, where clothe is rackt, as farre as threde will goe)
And whē the threede of wealth doeth breake, let wit and wisedom too
Doe what thei can to tie the threede, the knot will sure vndoo.
The storie treats no more thereof, yet therein maie you see,
That some haue vertues and good witte, and yet vnluckie bee.
In winnyng wealth, in worldly happs, whiche common are of kinde,
To all and yet the vse thereof, but to a fewe a sinde:
For some haue all their parents left, all thei them selues can catche,
And tenne mens liuyngs in one hande, and some haue nere a patche.
And some not borne to sixteene pence, finde twentie waies to get,
By happe yet some as wise as thei, no hande thereon maie sette:
I heard a white hoare heoded man, in this opinion dwell,
That witte with wealth, & hap with witte, would gree together wel.
But for to chuse the one alone, he held that happ was beste,
He saied witte was a happie gifte, but wealth made all the feaste:
Witte with the wise must companie keepe, then cold oft is his chere,
Wealth hath companions euery where, and bancketts all the yere.
Wealth hath the waie the cappe and knee, and twentie at his taile,
When witte hath nere a restyng place, no more then hath a Snaile:
Wit is compeld to be a slaue, to wealth and serue hym still,
Yet wealth is naked about witte, nought worthe where lacketh skill.
But if that wealth maie match with hap, then bid fine wit goe plea,
Our old Prouerbe is giuen me hap, and cast me in the Sea:
Unhappie must I iudge this man, in sondrie sorts and waies,
Yet fortunate I call hym then, in true report of praies.
The cheefest Iewell of our life, is vertues laude well won,
Whiche liu's within the other worlde, when fame of this is doen:

FINIS.

Churchyards

Churchyardes farewell from the Courte, the seconde yere of the Queenes Maiesties raigne.

Though Fortune casts me at her heele,
And lifts you vp vpon her wheele:
You ought not iope in my ill happe,
Nor at my harms, your hands to clapp.
For calmes maie come, and skies maie cleare,
And I maie chaunge, this mournyng cheare:
To gladsome thoughtes, and merrie looks,
Although you fishe, with golden hooks.
And make the worlde, bite at your baits,
And feede your selues, with sweete consaits:
Myne anglyng maie, at length amende,
My rodde it can, bothe bowe and bende.
As causes falls, for my behoofe,
I leaue you Courtiers in your ruffe:
I will goe liue, with plainer menne,
And vse my booke, and plie my penne.
Perhapps that I, asmutche haue seen,
As they that braues, it on the Spleen:
Where Cannon roard, and Dromme did sounde,
I did not learne, to daunce a rounde:
And baunte I maie, my happe the woorse,
I haue with many, a threede bare purse.
Been glad to serue, in Countries cause,
When you at home, were pickyng strawes:
Since you did spite, my doynges all,
And tosse from me, the tennis ball.
By woordes and woorks, and priuie nipps,
A man maie saie, beshrewe your lipps:
And vse a kinde, of ridyng Rime,
To sutche as wooll, not let me clime.

B.iij.   Where

Churchyardes Charge.

Where euery one, would Apples sheake,
Though at the hiest, the bowes are weake:
The Crowe bilds there, full saffe ye wotte,
And neare the topp, the fruite is gotte:
Well I full lowe, must beare my sailes,
In climyng often, footyng failes.
Watche you the ball, at first rebounde,
So I maie stande, on euen grounde:
And plaie at pleasure, when I please,
I am not greeued at your ease.
Although that you, with shiftyng braine,
Doe reape the profite of my paine:
And thrusts your hedds, tweene hap and nie,
Whose handes doe plucke, the barke from tree.
So greate and greedie is your gripe,
You eate the fruite, ere it be ripe:
And none maie feede, but you a lone,
You can not spare, a dogge a bone.
Ye cleaue together, so like Burres,
Perhapps in winnyng of the Spurres:
You maie the horse, and saddle lose,
When that her hedde, whose vertue flowes.
Shall see the deepnesse of your sleight,
And sette your crooked dealyngs streight:
And all your painted sheathes espie,
And waie what stuffe, in shadowes lye.
Thinke you she smiles not once a daie,
To see how many vices plaie:
Uppon the stage, where matter lacks,
You doe no soner tourne your backs,
But greater laughyng riseth there,
Then at the baityng of a Beare:
We thinke you thrist, your shopp not well,
In Court your follies for to sell.
That shopp stands full, within the winde,
Or els so muche, in peoples minde:

That

That if one fault be in your ware,
Teime thousande eyes, thereon doe stare.
And when thei finde, a counterfeite,
Or see, fine Merchaunts vse deceite:
Then crie a loude, wee smell a Ratte,
Some haue more witte, within their hatte,
Then in their hedde, that sells suche stuffe,
Well euery man, vnto his ruffe:
And I into, my coate of Frees,
For I in Courte, can hiue no Bees:
The hony there, is bought so deare,
I were as good, with countrey cheare.
Sitte free in mynde, and farre from stats,
And daiely matche, me with my mats:
As waite emong, the hautie breede,
Whose humourss are, full hard to feede.
Where smail is wonne, and mutche is spent,
And needlesse hands, doe stoppe the vent:
That well might serue, a thousands tourne,
Tushe at the pricke, to kicke and spourne.
I should but hurte, my shinnes ye knowe,
From Court to Countrey will I goe:
With mutche ill happ, and losse with all,
Now maie my boule, to vpas fall.
In alleps smothe, where it maie ronne,
I see in Court, shines not the Sonne:
But on a fewe, that Fortune liks,
And there a man, shall passe the Piks.
Eare he maie purchace that he craues,
As one doeth poole, an other shaues:
And marquesotts, the beard full trimme,
Yet nothyng runneth ore the brimme.
Till purse be full, and then perhapps,
When strings doe breake, there falles some strapps:
Into your hands, watche that who liste,
A birde is better sure in fiste.

Then

Churchyardes Charge.

Then fiue in feeld, keepe that thou haste,
Where wealth and witte, and tyme doeth waste:
Looke not to dwell, what drawes thee there,
But gaine or glorie, loue, or feare.
If gaine to Courte, doeth make thee goe,
Thou art no freend, but flatteryng foe:
That daiely seeks, thy self to helpe,
And couchest like the faunyng whelpe.
Till Prince hath filde, thy purse with pence,
And then sim subtill getts hym chence:
If thou in Courte, for glorie iette,
As dizard daunseth in a nette:
The worlde shall thee, rewarde with praise,
Was neuer Courtier in our daies.
So braue as he, then will thei saie,
And all not worthe, a trusse of haye:
At home thy loue, as well is seen,
And better, then in Courte I wene.
If like a subiecte, there thou liue,
And often good example giue:
To suche as stands thereof in neede,
If feare drawe thee, to Courte in deede.
The Prince can finde sutche quakyng soals,
She knowes whose harte is full of hoals:
And seeth what lucks in hollowe stocks,
And treads vpon sutche tremblyng blocks.
From sutche is bounties larges bard,
And then is bountie laced hard:
From suche the well hedde stopped is,
A volume could I write of this.
As large as any Chequer rowle,
But I the plaine, and sillie soule:
Must thinke and wishe the beste I maie,
And little of these matters saie.
Yet he that knwos, and giueth aine,
Maie iudge what shott doeth lose the game:

What

What shooter beats the marke in vaine,
Who shooteth faire, who shooteth plaine.
At little hoales, the daie is seen,
Some in this cace, maie ouer ween:
And thinke thei see in Milstones farre,
And take a Candle for a Starre.
Passe or sutche topes, and aunswere me,
What cause hast thou in Court to be:
If gaine ne glorie, feare nor loue,
To Courtyng doeth thy fancie moue.
What drawes thee thether hedlong now,
Giue eare, and I shall shewe thee how:
Thei sitte and stare in Courte some while,
Yea on the other doeth beguile.
With fairest semblaunce that is sure,
And euery craft, is put in vre:
To snatche or compasse that thei seeke,
Although it be not worthe a Leeke.
The finest hedds, haue furthest fatche,
The deepest sight, doeth neerest watche:
To trapp the vpright meanyng man,
And eche one doeth the beste he can.
To helpe hym self, by others harme,
These Courtiers haue so fine a charme:
I graunt there is honour wonne,
And thether ought the subiects ronne.
To shewe their dueties by some meane,
But why haue some consumed cleane:
Their liues and lands in this desire,
Ye knowe a man maie loue the fire.
Full well, and leape not in the flame,
Some thinke thei winne a goodly name:
When thei at home are Courtiers calde,
It is full gaie, if he be stalde.
An almes knight ere that all begon,
His happ is hard, that hopes thereon:

C.j.   Yet

Yet sith I sauour Courtyng well,
Would God I had moꝛe lands to sell.
To be at their commaundement still,
If that a man haue their good will:
He hath enough, what needeth moꝛe,
Old ladds maie ſhifte vpon the ſcoꝛe.
And let their garments ly and ſweate,
Oꝛ with their Oſtes wooꝛke a feate:
To ſette the hoꝛſe in ſtable free.
But now the wiues ſo hongrie bee.
And houſbands looke ſo nere their gaine,
A man as ſone on Salſbꝛie plaine:
Shall haue a cheate, as by that trade,
The daie hath bin, who could with blade.
And Buckler ſquare it in the ſtreets,
Had bin a minion fine foꝛ ſheets:
But now the pence doe make the place,
And woꝛlde is in an other cace.
Well let the matter paſſe a whyle,
And heare my tale, but doe not ſmile:
I hapt in Courte (as newe Bꝛoine maie,
That ſweepeth trimely foꝛ a daie.)
To be deſierd to plaie and ſyng,
And was full glad in euery thyng:
To pleaſe the Loꝛdes, and loꝛdely ſorte,
Foꝛ that ye knowe with chaunge of ſpoꝛte.
Theſe Courtiars humours ſhould be fedde,
And glad I was to bende my hedde:
And be at becke when thei did call,
In hope that ſomme good happ would fall.
To me foꝛ that apt will of myne,
Although my doyngs were not fine:
A Tabber with a Pipe full loude,
To better noyſe is but a cloude.)
Well as the Hackney is deſierd,
And ridden till the Iade betierd:

3

I did continewe long me thought,
And still I spent the small I brought,
And neuer got I one denere,
Then thought I to beginne the yere:
On Newe peres daie with some deuice,
And though that many men be nice,
And blushe to make an honest shifte,
I sent eche Lorde a Newe peres gifte:
Suche treasure as I had that tyme,
A laughyng verse, a merrie ryme.
Some thinke this is a crauyng guise,
Tushe holde your peace, world waxeth wise
A dulled horse that will not starre,
Must be remembred with a spurre:
And where there serues ne spurre nor wand,
A man must neds lead horse in hande.
So I was forste on causes greate,
To see in fire where laye the heate:
And warme their witts that cold did waxe,
But thrust the fire into the flax:
It will not burne if flaxe be wette,
The fishe these daies can shonne the nette.
And hide them in the weeds full ofte,
Thou knowest that waxe is tempered softe:
Against the fire, so frosen minds,
Must be assaied by many kinds.
To bryng them to a kindely thawe,
Who thrusts a candle in the strawe:
Shall make a blase, and raise a smoke,
An honest meane there is by cloke.
To sturre the noble harts from sleepe,
Whose coffers, custome makes to keepe:
Faste lockt, that should be opened wide,
To helpe the poore at euery tide.
Thei saie that knewe our elders well,
That often tymes thei hard them tell:

That larges linketh loue full faste,
And hardnesse loseth harts at laste.
And honour leanes on liberall waies,
And fame and honour nere decaies:
Till hoorde in horie mucke doeth holde,
The free and worthie vse of golde.
Oh sentence hye of Fathers wise,
I sweare by all the gods in Skies:
These woordes deserue immortall fame,
And nothyng is so mutche to blame.
As pintchyng hands that should be franke,
Admit the taker peelds no thanke:
To hym that giues, the gifte doeth binde,
Eche vertuous man and honest minde.
As captiue in all good respects,
To be a freende in full effects:
As farre as powre maie stretche vnto,
And thei that haue in warres to doo.
Can saie, what bountie bryngs about,
Where that is not, the fire goeth out:
And dyes as coale to ashes falls,
As Fouler taks the birde vp calls.
In strawyng corne and chaffe by heapes,
So bountie as a sickle reapes:
The harts and all within the brest,
No perfect loue can be possest.
Where franckenesse makes no place before,
Though force of earnest loue is more:
And lookes not on the gifte a whit,
It man in neede and daunger sit.
And finde their freends bothe cold and drye,
Then loue will shewe a lowryng eye:
And halte with you, as you with hym,
Although that some can cloke it trim.
I tell you loue is easly loste,
If you on loue bestowe no coste:

Thus

Thus as before I did rehearſe,
I ſent eche Lorde a merrie vearſe.
A iollie libell long and large,
And therein did good will diſcharge:
But nothyng did retourne to me,
That I could either feele or ſe.
Saue from a brooke, ſet yenne before,
Ranne dropps of gold, what will ye more:
Thus in this witred age of ours,
The ſmell is gone from goodly flowrs.
And golden worlde is tournd to braſſe,
Or hardneſſe dwells where beautie was:
There is no waie to gaine nor ſaue,
Then learne to keepe the thyngs we haue.
For he that wants ſhall hardly gette,
Except he fiſhe with finer nette:
Then either rime or reaſon knitts,
This worlde peelds not to pleaſaunt witts.
To baſeſt mynds ſometymes it bends,
For all the happs blinde Fortune ſends:
Doeth light on thoſe ſhe fauours mitche,
Some man you ſee can nere be ritche.
Though twentie yere he toyle and toſſe,
For he is borne to liue by loſſe:
And ſome that neuer taketh paine,
In worldly wealthe doeth ſtill remaine.
Ne Court nor Countrey ſeru's ſome man,
To thriue in, doe the beſt he can:
Then finde thou faut with none of bothe,
With blinde affection eche thyng gothe.
Happ lyes not in mans ronnyng ſtill,
Nor Fortune follows fineſt ſkill:
Nor he doeth not the wager win,
That in the race hath formoſte bin.
In Iudges mouthe the ſentence lyes,
So whether men doeth fall or ryſe:

Looke vp to hym that ruels the Skies,
The ritche the poore, the foole the wise,
And thei shall finde my woojds are true,
Thus foj a while, now Courte adue.

FINIS.

¶ Of a mightie greate personage.

Hen Thebus tooke his purple bedd,
   to rest from daies disease,
Maie seemde to dippe his golden hedde,
   vnder the Ocean seas:
And faire Lucina gonne to shine,
   and mount in starrie Skies,
Then crepte the sweete and kindely slepe,
a long my slombzyng eyes.

And pzickt me so to take a napp, that as in coutche I laie,
I dzeampt that Natures little babes, about my bedde gan plaie:
And bad me rise, and bewe a wojke, that kinde a newe would frame,
Foj that she thought bothe gods & men, would help to fojge the same
You speake but like yong girles qp I, she hath all ready boen,
Sutche wojks as now her hands would misse, if thei were vnbegon:
With that dame Nature had, I spide, with angrie visage redde,
And in her furie satte her doune, full right againgt my bedde.
Why foole quod she is Nature not, so perfite of her skill,
That she can giue to fleshe and fell, what shape and fozme she will:
Thou seest eche wozkman finer growes, eche wit doeth riper ware,
And knowledge can amende at full, the faults where cunnyng lacks.
The Goldsmith and the Caruer bothe, and all that wozks with toole,
Doe mende their hands and daely are, by Nature set to schoole:
The Pzinces pallace made of old, lookes like a sheepe coat now,
So if this tyme and Nature liste, to shewe their conyng thzow.
Wee can set fojthe a Candle blase, bepond the shinyng Sonne,
And take the light frõ twinkling starrs, whiles Moone her cours that
Can I not call foj Beautie whout, that I haue lent at large, (ron:
Haue not the hye immojtall Gods, giu'n Beautie to my charge,

                                  And

And maie not Nature breake eche mould, ý once her hand hath made,
And worke this yearthly drosse againe, vnto a finer trade:
Yes sure saied she, and I therewith, did humble pardon craue,
And at one instaunt by a signe, that mightie Nature gaue.
A thousande woorkmen all with tooles, came thrustyng in a rout,
And eche vnto his labour falls, as tourne doeth come about:
Thei blewe and puft and smoke out sweate, as though in thé did lýe,
To shape a mould, or shew through cloude, that *Venus* dropt from skie
Haue doen quod Kinde it shalbe thus, too long ye trifle here,
Then Cunnyng by her curious art, deuise suche collour clere:
That did the ruddie Rose disdaine, and passe the Lilie white,
If that a medley of those twaine, were made to please delite.
The woorkmen in this hastie broile, had raised vp a mould,
And eche one iu his office fine, had doen the beste he coulo:
Now satte thei still in silence sad, and rested for a space,
With that dame Nature by her skill, set forthe so trimme a face.
That Sonne and Moone and seuen starrs, did seem therein to shine,
In whiche the pleasant gods had plast, a paire of gladsome eyne:
Yea euery God one gift her gaue, as *Pallas* for her parte,
Possest her with a noble heade, to iudge or talke by arte.
And *Iuno* made request to *Ioue*, that *Venus* Queene of Loue,
Should neuer with false fonde desiers, her modest maners moue:
Dan *Cupid* brake a bowe for ioye, when this faire dame was made,
In signe ý she to *Dians* Nimphes, should walke in grenewood shade
The silly woorkmen seyng this, that seruaunts were to Kinde,
Trust vp their tooles and stole awaie, yet left the mould behinde:
Whiche as I gesse of diuers stones, was wrought by deepe deuice,
For therein Iazings might you see, and pearles of passyng prijce.
The Rubbie ritche, and prettie sparkes, of Diamonds clere & bright,
The Emerald greene, and Margarets faire, & Turkes blew to sight
Whose vertues passeth farre my penne, or yet my tong to tell,
Demaunde ye that of skilfull men, that knowes their Natures well
Loe foolishe man, loe here thou dolte, quod Kinde to me aloude,
How saiest thou is not this new worke, more faire then star in cloude
Doeth not this worke make all thé blusse, ý I haue wrought before,
Yea sure, for Nature is in minde, to make the like no more.

By

By this tyme was the Larke aloft, loude chirppng in the aire,
And eche one to their daiely toiles, gan busily repaire:
So rose I vp and rold in thought, where this faire wight doeth dwel,
And at the length I founde in deede, I knewe the worthy well.

FINIS.

¶ Of Beutie and Bountie.

When Beautie *Venus* doughter deare,
from *Ioue* descended downe,
To reigne on yearth an Empresse here,
with scepture and with Croune:
To Pleasures pallace she repairde,
where with a Princelp porte,
She helde an open housholde long, in feasts and royall sporte.
The fame whereof rang through the worlde, so shrill in euery eare,
That well was him, & glad was she, that might come banquet there:
The lifts were made, the scaffolde deckt, eche thyng in good arraie,
The Lords full braue, the Ladies fine, the Courtiers trim and gaie.
And as these states in triumphe were, all placte in their degrees,
And to beholde the shiuerd staues, the people swarmde like Bees:
In stept a goodly armed knight, on courser white as Snowe,
And twise he paste the Tilte about, as soft as horse could goe.
And when he came where Beautie satte, he pausde with bowed hed,
And loude in open audience then, all haile faire Queene he sed:
I came quod he from Manhoods court, the worthiest prince aliue,
Who keepes his kyngdome all by sworde, and doeth for honor striue.
By battaill and by breakyng launce, who sent me hether plaine,
To chalenge for my mistresse sake, the stoutest in thy traine:
No soner he his message saied, but in there rusht a bande,
Whose clattering harnesse causde their steeds vpon no groud to stad.
The dust flewe vp, the preace did shrinke, the fompng horses naied,
The trumpets blewe, the launce in rest, the spurres on sids thei laied:
Fie cowarde knight quod Courage then, can all you fight with one,
So thei retierd, and to the shocke, came youth all armde alone.
These champions met as yearth should shake, so fierce thei seemd to be

As

As man became a Lyon woode, and horse in aire should flie:
At eche encounter crasht their staues, and fell amid the throng,
The buffetts were so freely dealt, the blood through Beauer sprong.
The Queene cride hola, cease quod she, you turne your sport to spite.
Some cause your collour doeth encrease, & Mars the pastime quite:
A cause quod Youth (moste worthy dame) and my leege Ladie dere,
Came euer yet before a Prince, so stoute a chalenge here.
Who dare with *Venus* doughter boste, dame Beautie iustly calde,
That came from Skies, and satt next *Ioue*, in sacred honor shalde:
Though Beautie sprang frō yearthly cause, & had but shape of kinde,
And did no heauenly gifts possesse, nor vertues lodge in minde.
Yet Boldnesse churlishe chalenge braue, too sawsie is you knowe,
And Beautie hath too many freends, to see her handled so:
When Boldnesse hard this taunting tale, & markt the peoples chere,
He thrusted through the thickest throng, and drewe the scaffolde nere
And all on knees he crau'd to speake, and aunswere to this race,
On whom the Queene for honours sake, did shewe a gracious face:
Speake on quod she, so stept he vp, and thus to her he saied,
O puissaunt prince, thinks Youth of braggs, y boldnes y ands afraied
I am a braunche of Manhoods blood, that stoute conceite begate,
The hope and helpe of hie attempts, and staie of euery state.
That hether came for that no Courte, can be where I am not,
No Tornay seen, no triumph made, no fame nor glorie got:
And wotte you well, a Princesse too, in Court I serue this howre,
That is as greate in some respects, as she is small in powre.
If stately honour can be gest, by goodly graces trime,
Or perfect beautie be possest, where Bountie swimes at brime:
Or wisedome vnder seemly shaeds, maie shine or yet be seene,
My mistres is a worthie dame, though Beautie be a Queene.
Report hath blowne to Manhoods eares, the troeth of that I tell,
Then Boldnesse needs not blushe to boast, y Bountie beares the bell
And sith you licence me to speake, I dare deuoide of blame,
Light suche a torche vnto your eyes, shall shewe this Ladies name:
When Skie is clere, and Sommer set, to shewe the weather faire,
I meane when calmie blowes the winde, and pleasaunt is the aire.
A Marie gold then maie you finde, full nere an Eglantine,

D.J.   Whose

Whose flowrs within the Porch new buds, & yet in court both shine:
Her countenaunce carries sutche a state, full right amid her face,
As though therein the Muses nine, had made their mansion place.
A ratlyng sounde vnto your eares, of her now here I showe,
Now racke & wrest my meanyng out, and you my mind shal knowe:
This saied eche one on others lookt, and he on horsebacke leapt,
And some that dwelt in their concept, full close in corners creapt.
The glorious sort that gapte for fame, where no deserts could be,
Did drawe a backe and preast a pace, with plaine reproche to flee:
The hautie minds held downe their heds, hye looks gan blush for fere,
As youth beheld this sodaine chaunge, he thought no taryyng there,
The Gods regardyng from the starres, what strife by Beautie rose,
Bad Venus call her daughter home, and homewards so she goes:
Then sawe I Boldnesse turne againe, who gaue for Bouties weare,
A garlande of the goodliest flowres, that euer pearth did beare.
And foarst her for to take the same, in signe of glorie wonne,
As Beautie mounted to the Gods, and all the triumphe doen:
The people seyng Beautie gon, with one assent did crie,
That Bountie pleased more their mindes, then Beautie did the eye.

FINIS.

¶ Of one that by dissemblyng,
fedde his desire.

IF loue be luste, the more my latke,
and lesse I thinke your lucke,
Yet loue I not for leude delight,
nor gaine of worldly mucke:
But for a finer freake, be you the iudge thereof,
When craft to cloke some secret smart,
begins to scorne and scoffe.

Witte workes with words and wiells, a waie to winne his will,
And where y sleight shewes gladsom smiles, y world conceius none ill
Mirthe blears the peoples eyes, and makes the matter light,
And sadnesse breeds suspect to sone, in heds of deepe foresight.
And worlde mislikes no ropes, that mirrie laughter brynges,
God knowes what care the bird doeth feele, in cage that swetly sings
Some

Some weepe in wedoyng weeds, and laugh in mournyng gownes,
And sure I smile my self sometyme, when froward fortune frounes.
Where is moste cause of care, moste signe of ioye I showe,
For pleasure is redoubled oft, where men dissemble woe:
Who bluntly bites a baite, and swallows vp a hooke,
Is caught like Gogon in a nette, or conquerd by a looke.
But sutche as warely feedes, and pikes out bones still cleane,
Shall eate their fill, & learne to knowe, what daintie morsells meane
Thus restyng at your will, I feede my hidden thought,
With fancies merrie sweete conceipts, a foode full dearly bought.

FINIS.

¶ Of stedfastnesse and constancie.

W[hen] Constance maks, her bed in bloudie breast,
And builds her bowre, with bowes of bloming trothe:
There frendly faithe, is sure a welcome geast,
And Ioue doeth dwell, and Ladie Venus bothe.
The Gods are glad, to vewe sutche trothe belowe,
The heauens hopp. to see sutche Constance flowe.

But where fonde luste, doeth leade firme loue awry,
And fickle toies, in feeble fancie falls:
And foule delite, doeth feede the wantons eye,
And stedfast harts, are toste like Tennis balls.
There Pluto raignes, with all his hounds of hell,
In irksome shame, and smoothryng smoke to dwell.

Oh what a praise, hath Constance shinyng face,
What greater blott, maie be then breache of loue:
The constant minde, hath sodaine change in chace,
But thei that will, of eury water proue.
Shall drinke sowre whey, in steede of sirup sweete,
For licrus lusts, a licour fitte and meete.

Tenne thousande false, I finde where one is true,
With faithe forsworne, loe eury face apears:

These

Thefe faithleffe fooles, that chainge fo2 eurp newe,
Doe looke full fmothe, pet p2oue but fcrattpng B2ears,
Since foule deceipts, hath filde the wo2lde with bice.
Wle ought to giue, dame Conftance all the p2ice.

   O blafpng ftarre, that burnes like *Eathna* flamne,
O fickle daines, goe hide pour hedds in holes:
App2oche not nere, where J doe Conftance name,
Pour dwellpngs are, emong the dampned foles.
Goe girmpng girles, and giglotts where pe lufte,
Dame Conftauce fits, in glo2ie with the iufte.
       FINIS.

   ¶ *Of one that founde falfhed in felowfhip.*

IF faithe take foile, and plaine good will be lofte,
  Let fained loue, feke Larks when Skie doeth fall:
  Jf triall greate, be made a double pofte,
  No p2actife feru's, to fhoffull Cards with all.
  Jf waitpng long, can winne but cold reward,
Bid wilie witts, goe warme his hands at fire:
Jf trothe want, happ, fo2 toile and greate regade,
There is no hope, that wo2keman fhall haue hire.
Jf letters large, but little likpng winne,
Pour bablpng tongs, in fine fmall bofte fhall make:
Jf feruice pafte, a fute muft newe beginne,
Newe hangers on, in hafte their leaue maie take.
Since fuertie fh2inks, and freendfhip fmells of gile,
Adue badd wo2lde, thp fauour lafts no while.
      FINIS.

   VVritten to a vertuous gentle woman,
     whofe name is in the verfes.

DEme all my dedes bp true defarts,
   that fheweth eurp frute,
And paife mp woo2ds, and p2oue mp woo2ks,
   and fo efteme my fuite:

M y trothe vntried bids me retire, and bꝛyngs me in dispaire,
P asse on saith hope, good hap maie come, the weather maie be faire.
P rease not to faste saicth Danger then, foꝛ feare thy foote doe slide,
O f hastie speede greate harmes doe rise, as often hath bin tried:
R epentance comes care men beware, foꝛ want of perfite skill,
T herefoꝛe let reason rule the raine, and wisedome maister will.

Thus in myne hedde a battaill is, betwene my hope and dꝛeed,
Hope pꝛicks me foꝛthe, feare dꝛiu's me backe, my fancie thus I feed:
Though hope be farre aboue my happ, good lucke maie me aduance,
And this great warre maie be a peace, as al things haue their chance.
The tossed shipp maie hauen it, that anker holde hath none,
As rainie dꝛopps by length of tyme, maie pearce the Marble stone:
What foꝛt oꝛ holde is halfe so strong, that euer man could make,
But poulders foꝛce and Cannon blast, can make it doune to shake.
The pelletts all that I must bꝛyng, vnfained faithe must be,
The ladder foꝛ to scale the walls, is trothe when tried is he:
This aunswere maie the captaine make, to whom my siege I laie,
Whose foꝛt is wonne by sutche a fault, oꝛ by none other waie.
With Ensigne spꝛed, and battrie set, I hope to make a bꝛeache,
And trust to winne by suite at length, that now is past my reache.

FINIS.

A farewell to a fondlyng.

THe heate is past, that did me fret,
The fire is out, that Nature wꝛought:
The plants of loue, whiche youth did set,
Are dꝛie and dedde, within my thought.
The Froste hath kilde, the kindly sappe,
Whiche kept the harte, in liuely state:
The sodaine stoꝛmes, and thonder clappe,
Hath tourned loue, to moꝛtall hate.

The miste is gone, that bleard myne eyes,
The lowꝛyng clouds, I see appere:
Although the blinde, eats many flies,
I would she knewe, my sight is clere.

Churchyardes Charge.

Her sweete diseuipng flattryng face,
Did make me thinke, the Crowe was white:
I muse how she, had sutche a grace,
To seeme a Hauke, and be a Kite.

Finis.

¶Written to the good Lorde Maior (of London now in office) called Sir Nicholas Woodroffe Knight.

He tyme showes all, as fire woorks ware,
  in tyme greate thyngs are doen,
Tyme weau's the web, and wrought the flaxe,
  that paine through tyme hath sponne:
Tyme must be sought, tyme must be vsde,
  tyme must be tempred well,
Els out of tyme, in any sorte,
  the tale is that we tell.

So tyme moues pen, & sturrs the muse, (that time had lulld a slepe,)
To write of tyme and matter sutche, as maie good credite kepe:
Then my good Lorde, to former tyme, I doe referre my verse,
And auncient yeres, with elders daies, that can great things reherse.
Tyme brought the sworde (that eche one fears) to rule the rurall sort,
Tyme wanne this Citie hye renowne, and gatt it good report:
Time made the chosen Maior a knight, and time did greater things,
For tyme made subiects loue the lawe, and honour rightfull Kyngs.
Thus tyme was nours, and mother bothe, to chosen children here,
And tyme out worne, takes life of trothe, so showes like candle clere.
Whiche time my verse reuiu's againe, and bringeth freshe to minde,
The tyme that long is paste before, and thousandes left behinde:
For those that in this present tyme, list looke on Elders daies,
Who in their tyme did some good deeds, and reaped peoples praise.
As gwerdon for the tyme well spent, and vertues right reward,
That giuen is to graffs of grace, that God doeth mutche regard:
As tyme hath taught, good men to rule, and made the bad obaie,
So tyme hath rooted vp all weedes, that made good flowers decaie.

This

This Citie claimes by tracte of tyme, a stately Ciuill trade,
And is a Lampe, oʒ ſhinyng Sunne, to Countries ſillie ſhade:
Foʒ Ciuill maners here began, and Oʒder roote did take,
Whē ſauage ſwaines in rubbiſhe ſoiles, did ciuill life foʒſake. (tends,
Here wit thʒowe wiſedome weldeth wealth, ⁊ woʒlde good tyme at=
And God though trafficks toile ⁊ paine, a woʒlde of treaſure ſends:
Here ſtates repaire, and lawes are tried, and noble cuſtomes ſhine,
Here dwells the Sages of the woʒlde, and all the Muſes nine.
The Court it ſelf, ⁊ Innes of court (where wit ⁊ knowledge floes,)
Haunts here as terme and time cōmands, and people comes ⁊ goes:
Here are Embaſſours feaſted ſtill, and foʒaine kynges haue bin,
Here are the wheeles of publike ſtate, that bʒyngs the pagent in.
And here is now the Maiden toune, that keepes her ſelf ſo cleane,
That none can touche, noʒ ſtaine in trothe, by any cauſe oʒ meane.
Then here ought be no member left, that maie infecte the reſte,
Whip faultoʒs hence, and plage the woʒſt, and make but of the beſte:
Let ſtubburne route be taught to woʒke, bid paltrars packe awaie,
Giue Idell folke no lodgyng here, cauſe wantons leaue their plaie.
Searche out the haunts of noughtie men, ⁊ bʒeak the neſt of theues,
Yea plucke their liurep oer their eares, and badges from their ſleues:
That bʒeeds miſrule, and rudeneſſe ſhowes, ſo ſhall the Ciuill ſeate,
(As Lanterne to all Bʒitaine lande) remaine in honour greate.
Demaūde how thʒedebare figboies liue, ⁊ ſwearing dāpned ſpʒetes,
Refoʒme thoſe blading deſpʒate dicks, that roiſte aboute the ſtretes:
Diſperſe that wicked ſhameleſſe ſwarme, that cares not foʒ repʒoch,
Purge eury houſe from graceleſſe geaſtes, that ſetts all vice abʒoche.
Rebuke thoſe common alehouſe knights, ẏ ſpends awaie their thʒift,
And aſke on Benche where Juſtice ſitts, how roges ⁊ beggers ſhift:
Teache railyng tongs to tune their ſpeeche, and talke of that is fitte,
Holde in the raſhe and harebʒaine heads, by Lawe and Oʒders bitte.
Knowe whence theſe fauſie libells come, ẏ faine diſcoʒd would make,
And wooʒke by art and crafte to pluke, the ſtyng from ſubtil Snake:
This Citie is no harbʒyng place, foʒ veſſells fraught with vice,
Here is the ſoile and ſeate of kyngs, and place of pʒecious pʒice.
Here woʒthies makes their mantions ſtill, ⁊ builoeth ſtately towers
Here ſitts the Nobles of the realme, in golden halles and bowers:

D

Churchyardes Charge.

O London looke to thy renowne, thy fame hath stretched farre,
Thou art a stare in tyme of peace, a helpe in cause of warre.
A feare to foes, a ioye to freends, a Iewell in our daies,
That well maie matche with any Towne, or seate of greatest praise:
Here people are so meeke and milde, that forraine nations throwe,
In Ciuill sort, with wealth and ease, maie liue in quiet nowe.
What Citie can make boste and saie, (great God be blest therfore)
It doeth so many straungers feede, and so maintaine the store:
For here the more the number is, the lesse of want we finde,
Of corne and cates, suche store is here, it answers eche mans minde.
Waye well the dearth of other realmes, and you shall see in deede,
The plentie of this litle Ile, supplie our neighbours neede:
In worlde who trauailes any where, and then repaireth here,
Shall finde eche thing good chepe at home, that is abroade full dere.
And none but London note it well, doeth keepe one stint and rate,
Of victailes in the market place, looke throughout euery state:
Yea, here when God for wicked life, his bountie will withdrawe,
The Maior and brethren shonneth dearth, by rule and noble lawe.
Here is prouision for the poore, and who that markes the same,
Shall see that worthie Sages graue, deserues a noble name:
My boldnesse now (O my good lorde,) excuse through my good will,
That euer in my Countries praise, is prest and readie still.
And where the naughtie liues of some, are touched by my penne,
It is for Londons honour spoke, that can reforme suche menne:
Whiche in this stately shepheards foide, like rotten shepe doe liue,
And who for want of lookyng too, doe ill example giue.
God graunt whiles worthie *Woodroffe* rules, (& euery other yere,
There comes no Mothes emong good men, nor Caterpillars here:
Thus wishyng well, in Londons laude, my penne I must excuse,
To Printer sent these verses plaine, of this laste mornyngs muse.

FINIS.

www.ingramcontent.com/pod-product-compliance
Lightning Source LLC
Chambersburg PA
CBHW021429090426
42739CB00009B/1410